The Academy of Hay

JULIA SHIPLEY

Winner of the 2014 Melissa Lanitis Gregory Poetry Prize

Bona Fide Books
Tahoe Paradise, CA

Copyright © 2015 by Julia Shipley.

All rights reserved. No portion of this work may be reproduced or transmitted in any form or by any means, electronic or mechanical, including photocopying and recording, or by any information storage or retrieval system, without permission in writing from Bona Fide Books.

ISBN 978-1-936511-15-0

Library of Congress Control Number: 2015945068
Copy Editor: Mary Cook
Printing and Binding: Bookmobile
Cover image: Pictureguy/Shutterstock.com

Orders, inquiries, and correspondence should be addressed to:
    Bona Fide Books
    PO Box 550278, South Lake Tahoe, CA 96155
    (530) 573-1513
    www.bonafidebooks.com

Acknowledgments

Grateful acknowledgment is made to the editors of the journals and anthologies who first published the following poems. The poems, sometimes in earlier versions, appeared as follows:

Bloodroot: "The Heifer" and "Jewelweed"
Burnside Review: "In the Far Field and the Near"
Cincinnati Review: "The Needle" and "On This Day in History"
Colorado Review: "The Lasts"
CutBank: "Field Work" and "Part of an Argument"
FIELD: "Timothy" and "Field Revealed as Runway by Morning"
Flyway: "Bird Count"
Gihon River Review: "One Wind"
Green Mountains Review: "On the Road"
Matter Press: Journal of Compressed Creative Arts: "Heartacre"
Northern Woodlands: "Persuasion"
Pierogi Press: "Horn"
Poetry East: "Being"
Rivendell: "Heron, Gnomon"
North American Review: "Migration of Baling Twine"
The Salon: "Two Eggs" and "Noah and Persephone Raveling"
Taproot: "(G)loves"
Terrain: "First Do No Harm" and "o"
Vermont Life Magazine: "Twine"

The following poems appear in *Herd*, a chapbook published by Sheltering Pines Press in 2010:

"Winter as a Balance Sheet," "Twine," "The Present," "The Heifer," "Horn," "The Let Down," and "Every Year."

The following poems appear in *Planet Jr.*, a chapbook published by Flyway/Iowa State in 2012:

"Heron, Gnomon," "Our Mistake," "One Wind," "Bird Count," "Field Revealed . . . ," "Cornfield," "o," "Timothy," "First Do No Harm,"

"Being," "Two Eggs," "Persuasion," "Elegy for Anger," "The Heifer," "Kokoro," "Every Year," "Winter as a Balance Sheet," and "Gust of Flowers."

TABLE OF CONTENTS

I. Herd

Narcissus Cleaning the Bulk Tank / 1
Horn / 2
The Day We Woke Up With and the Day It Is Now / 3
Jewelweed / 4
Bird Count / 5
Heron, Gnomon / 6
Ballistics / 7
Spoils / 8
Being / 9
Ghost /10

II. Barn Storms

The Garden of Whirligigs / 13
Our Mistake / 14
Heartacre / 15
Draft / 16
After the September Storm / 18
Persuasion / 19
Noah and Persephone, Raveling / 20
Elegy for Anger / 21
Cornfield / 22
Every Year / 23
The Letdown / 24
On the Road / 25

III. The Academy of Hay

Migration of Baling Twine / 31
Part of an Argument / 33
Field Revealed as Runway by Morning / 34
Hapless / 35

Winter as a Profit and Loss Statement / 36
Notes About Snow / 38
One Wind / 39
The Present / 40
The Christmas Season / 41

IV. Husbandries

Her/Herd / 45
Discovering Venus, April 1820 / 46
On This Day in History / 47
First Do No Harm / 48
In the Far Field and the Near / 49
The Lasts / 50
Gust of Flowers Over Us / 51
Kokoro / 52
Eating / 53
Forms of Worship / 54
Field Work / 55

V. Herd of the World

Sufficiency / 59
Beforemaths, Aftermaths, and Polymaths / 60
o / 61
(G)loves / 62
The Needle / 64
Twine / 65
The Heifer / 66
Two Eggs / 67
Nine Miners / 68
Mappa Mundi / 70
The Holding Pen / 71
An Exegesis / 72
Timothy / 73

I. Herd

Narcissus Cleaning the Bulk Tank

She assembles
under my industrious hand:
worried out of occlusion,
beneath my agitated rag,
she grins back from the polished steel.

When I reach for her,
she welcomes me;
if I recoil
she'll flee.

Horn

Imagine growing something from your head
that could blind you: She once knew a ram whose horn grew askew,
toward his eye. Slowly, though, as slow as hair grows,
as gradually as summer spears through winter;
as long as it takes something newly known
to be said out loud—

tense as when you watch a wine glass set
on the table edge
beside a guest telling dramatic stories with his hands . . .
The morning the horn grazed eyeshine, at the brink of his blindness,
the farmer showed up, hacksaw in hand.
She keeps this stub to remind her.

The Day We Woke Up With and the Day It Is Now

My turn to lure the sheep out of the garden;
in my nightgown and muck boots, I fill their rubber bucket
with two quick scoops of feed.

The sky is still hooded with the same dark as the barn.
I traipse halfway out between where I want them
and where they are: waiting, a subtle pursuit.

I know the one who found the gate incompletely latched
and nudged it till the hinge splayed, and let the others
straggle in to graze on tender lettuce, bean shoots, coils of peas.

Morning starts with a wily increment,
and soon the whole flock overtakes night.
Now I try to entice that one sheep,

shaking my bucket of grain inside her sight,
while the mountains settle out of the plum sky
I'll follow all day.

Jewelweed

Needs only dry weather and merest pressure
to split open and launch seed;
I wish you'd burst beneath
my touch.

All week I've watched purses of milkweed
leak a soft exhaust—
fine hairs climb successive breezes,
still you resist.

I've wanted you to open like a ribbon with a slipknot.
But whose mere stroke could coax me
to spring to light
alone.

Bird Count

they feel the need to steal what is freely given—overheard

I.

When the loon launches its dire
lutish cry, the notes pelt my skin,
the sky shatters and the pond dimples in rain.

II.

Porter, a toddler, hands me a turkey feather, *Here*.
He gives me a thing, but uses the language of location.
I jab my heaviest thoughts with his feather.

III.

The hen uses her beak to nudge-pull
the egg under her breast,
like a man tucks a pack of cigarettes in his shirt pocket;
I took chocolate from the checkout shelf when I was Porter's age.

IV.

Now I venture into the loon-cried rains,
to raid what the chicken gave.

Heron, Gnomon

I.
But it lingered, I fled.
It stood still as a needle
touched to a phonograph,
while the water slipped through
its simple legs, watched me
pass threadlike through the opening day.

II.
An Eskimo culture offers an angry person
release by walking the emotion out of his
or her system in a straight line across the landscape:
the point at which the anger is expressed
is marked with a stick, bearing witness to the strength
or length of the rage.

III.
Once I suddenly noticed
I had no shadow except
directly beneath me. I
straddled a black so opaque,
it was the hole and the earth
was an urn
into which I will fit
entirely.

Ballistics

It came back—brass-capped cylinder, impotent amid the driveway's acorns and gravel. Plastic casing printed with: *Federal* and *Heavy Game*—my housemate's, no doubt, knocked out of his truck by accident, which is what hunters try not to have. They try for intentional death. What do I do with his almost innocent, what my hand is too small to hide entirely: the brass end glints.

*

Once upon a time I thought, well, eighteen's long enough, and I sighted a skinny dishwasher, Bullet, who already knew how to shoot. Consensual, yes, but not much of a romance; would I if his name were Eric? From the inside: bits splattering into tissue, game—like the stakes aren't serious. It didn't register, by a long shot; I moved away. Eighteen years later I remember his acne, his goofy incredulity, *You sure?*—these embedded.
Time's a kind of magazine, heat seeking, detail smearing, just like a drink, being loaded and loose, keeping *fast* infused with *free*; and then one day it hits: he was sixteen?

*

This just in: a bullet can fly two miles from the barrel. Here's this man, let's say it's Eric, sitting in front of the TV when a stray finds his spinal column, done. Whereas that hunter may have no idea where his ammo sped; almost innocent. Any poem can travel further, though its wound is inconclusive, weapon of mass compassion, *Hold it, filly*—
Who calculates the impact? Aiming as I am, I am now.
No more hunting. Only spree.

Spoils

> *Your Majesty, that we human beings should be made of limp wet meat appeared to me as strange as that we should be also air and spirit.*
>
> —Haniel Long, *The Power Within Us: Cabeza De Vaca's Relation of His Journey from Florida to the Pacific, 1528–1536*

Her innards weigh as much as Tony, the butcher,

who saved her digestive tract in a fifty-gallon barrel for the purpose of our study:

what (grass) becomes

—here—palate, esophagus, rumen, reticulum, omasum, abomasum, large intestine, small intestine, anus:

grey slick, and pink as newborn, notice the honeycomb structure
of the second stomach, shag-carpet texture of fourth, brown juice leaking ...

and sheathed in green flies, possibly

five hundred, humming, shimmering, a curtain of green sequins,

lifting, reclinging to the guts of this cow,

attraction adheres, in spasms—fly-like: *how much can we take?*—

releasing, and reattaching to this exposed system of change:

the place we eat and love between

Being

I mean the bumblebees
with no allegiances,
except to self and burrow (as body to shadow),
shuttling from mallow to mallow,
to flummox those pink–sided stadiums, *a revelry*,
tussling with stamen, stoking gold on their abdomens,
till each rises, buzzing, a thumbnail sketch
of heaven at midnight: velvet black thorax
filthy with outer galaxies.

Ghost

Twilight was sticky, damp, almost rotten.

He shucked off his shoes and left them by the front door.

In the morning the floorboards, painted the color of dried blood, bore his white footprints.

The talcum track of his long, skinny foot crossed the threshold into the living room.

For a few days, I sidestepped them.

Then I slashed at them, swept and swabbed every vagrant trace.

II. Barn Storms

The Garden of Whirligigs

Yellow leaves fleck across rock—
butter-colored ingredient of green
shows now and sun hunts the sky's far side.

A real duck squawks above
my garden of whirligigs where
everyone goes nowhere
but spends wind trying:
spoof of a Canada goose whirling
its furious wings, futile—

I took my spackling knife and scraped
the wings of living moths out of wet paint.
In spring I'll find the road pockmarked
with little caves, their stones all rolled aside.

Our Mistake

In the black grass and bilge-black night,

before the full moon peers into our wrack,

crickets issue a frantic dispatch

from the offices of black hostas,

in the final ticking seconds

of our black euphoria, in the burnt faces

of sticky black peonies . . .

Heartacre

The farmer squats in the dirt road beside the tractor. Using a twig he draws a way to enter the field. For the next two hours I'll drive twenty-five miles, mowing circles within circles. On the pond I'd watch ripples erupt and vanish, thinking: *fish postulants*. Sister Bernard Mary once told me, "A journey that begins with him leads to him." It's true of the ryegrass with its three miles of root hairs per day; 6,603 miles per season, traveling all that way, and going nowhere. We used to make fun of something called the "corn joint," the lignin keeping the stalk upright, stationed despite wind. The farmer, stoic mostly, suddenly spoke:

Could you stand in a field for three months?

Ah, but our potato fork did. It remained through the winter, an effigy for where we did stand and bend, and drive around with the cultivator all those months, between sowing and harvesting. Yeatsian, a *tattered coat upon a stick*, he grows more like his scarecrow, hair amuck, grease on his pants' hip when he's pinned to his tractor, haying. That tilted fork still stuck in February, buried to the grip, represents everything: his investment, and my investment, our stubborn stab in this endeavor, our vulnerability to the elements, well, I'll speak for me, for whatever comes or falls or fails, I'm steeled.

Draft

It's not the team I expected to last
hitched to a stone boat burdened
with thirteen blocks of concrete:
each block five hundred pounds
and the boat itself (picture a sled) another K.

The team, considered "light,"
meaning anything under 2,800 pounds,
needs three men: one on either end
of the whiffletree and a middleman tight on the reins.

A clang, their harness hook catches the boat's ring,
and the horses surge—*Hyeah!*—
budging the boat's dumb load
from standstill to slow go,
which jerks a slack rope tied to the boat's hind,

unspooling it the regulation six feet—
till the rope goes taut, topples its jug of sand,
 —now the pull is done: a minor journey.
The spectators applaud, a Good Pull—
we saw them struggle under half as much,
yet they're still here, long after other
drivers waved to the judge, a quit.

A Bad Hitch is when the team pulls unequally
as we have, and the mismatched lurch
keeps the boat at a halt, despite their heart.

When horses pull together
they pull five times their entire weight,
though "two horsepower"
is supposedly all they contain.

It's more than muscle, more
than they should be capable of,
and why, by god, for the pot-bellied
guy with the reins, do they want to
pull so hard, as we did in every
event, mistaking freight for fate?

ter the September Storm

Amid humidity, low mosquitoes, tomatoes baubled along the sill,
I render this because the roiled creek's gone back to silk,
the wood thrush is suddenly mum, a buzzard's dizzy over skunk guts,
and with ill-attended ceremony or none:
this signals an end.

The lightning bug's drubbed in the water bucket; ducks jeer *hahaha*
toward four o'clock; inelegantly and with dispersed fanfare,
as one season closes here, flagged by tattered cosmos,
another opens:

both harpoon and shepherd's crook—a tall stalk stuck in the lawn gone belly up—
pink hollyhock gawks at an oncoming frost.

Persuasion

The saw persuades the tree to forfeit an upright
position and relax its length against the earth.

The hammer persuades the washer to lie flat in the hole
the drill has persuaded the wood to open.

What we need is some persuasion
to send the tenon home.

I pulled up a post in one dream
and pounded it into another. I dreamt a frame:

wood curls, bunged nails, extension cord
coils on the floor; nearby more planks of wood

while you stood through the unfleshed truss
exalting the use of force.

Noah and Persephone, Raveling

November, Noah stands in the bed of his one-ton dump truck
and lays in the boards one upon the other, so he disappears in
increments: quarter gone, half gone, three-quarters . . .
when the back of his truck is walled up he calls out, *Sweetheart?*
as if she has left him.

April, and after she does, she watches another Noah stride to the
back of his delivery truck and yank up the door. There, in the
dark hold: only branches of cherry blossoms—their tangled legs
and wrists, their faces blind luminous blossoms.

Elegy for Anger

Up where harvested garlic hangs with mud-caked buttocks
curing, where flies duel the wasps,

she's crouched on an upside-down milk crate, with a knife, clippers,
whittling the dirty wrappers off garlic,

snipping the stems and root fringe,
pitching each finished head in the basket.

Fog blots out the ground below.

A cricket grates a hymn from the heap of garlic skins,
a pile so light any breeze could dishevel them.

Cornfield

How I miss the cornfield, the lap of snow it held,
how it was cardboard-colored mud,
how they treated it like cardboard, crisscrossing,
compacting, dosing it with ammonium nitrate,
genetically modified seed, spritzing it with atrazine mist,
how I miss the tall stalks falling in the maw of his chopper,
and the shreds snowing from the backs
of overloaded trucks barreling back to the bunker,
how I miss the geese settling in for a caucus
and the two boys skulking down with guns
as long as their attentions spans.
Most of all I miss the rainstorms we heard breathing in the corn leaves,
heard as wind and water pelting their ragged belts,
crossing from far field toward us,
heard as it approached, heard as it arrived, then felt,
pecking at our hair and shoulders,
as if we were the last stalks.

Every Year

The chopper devours eighty acres
and makes confetti,
takes all the field gave except
a single corn stalk,
its calligraphic leaves askew,
tassel slanted—
an ideogram for the word
auspicious.
The wind in its papers scratches
the stunned afternoon.
The chopper's raucous blare is distant, dimmed
to the size of a fly's buzz—

a sound I want to crush with my boot,
but can't.

The Letdown

He looks pained
by certain words
said aloud.
Aloud is too loud.
He winces as if
the worst were screamed.
He keeps a herd
of words inside—
they mill around
like cows in a free-stall
barn at night,
their full udders ache.
His lips purse
for a kiss.

On the Road

When I say I'm *perturbed*, I mean—
I'm casting slurs, hurling curses
like splats of floury lime dropped behind fat-tired trucks
leaving a trail of powdered flatulence.

I mean I'm hulking comments like the goddamned
hoodlums who throw tires over the embankment
and now they loll deep in the ravine,
like diabolic Cheerios;
I can, I have thrown things, I know, but

I'm distraught, yes, acting like a jackknifed Mack,
my seven-ton payload of grapes
becoming juice for every ant in Marin.
Or a capsized armored truck spewing
two million in cash near Linden's exit 12, but, sweetheart:
you will never ever have to snow shovel my dimes.

By upset I mean something—
oil, mercury, carbolic acid,
hydrochloric acid, methyl parathion,
liquid tar—
is beyond my containment, yes.

I am volatile where volume
for storage would be far more ideal

I concede: at times my equilibrium
has the same perspicacity and velocity
as the tractor trailer in Ypsilanti—and oops,
there go my hundreds of cartons of metaphoric eggs
cracked on tarmac

justifying anyone's "I feel like I'm walking on eggshells."
This is your point, right?
Because, of course, I *have* become unhinged, but not in excess
of fifty thousand screws lost on the thruway.
Just one screw loose. Or lost.

*

Once there was a thirteen-ton dump truck that plowed into traffic,
which, on impact, sent its dirt payload skyward—

Say I'm not like that.

At worst I'd strew soot over every
rube/sob/stooge/pawn/dude who ever asked,
How are you?

All you strangers suffused in my answers: here is my retroactive amends,
dustpan and brush in hand:

I did strew dimes, limes, molasses, carcasses
all because he did not call.

This is also true: Once, thereafter,
I was dispatched to deliver a dolly full of milk bottles
to the customer's hatchback when the wheel stuttered and the weight
shifted and the crate pitched headlong
and all six bottles broke.
As the cool blood of milk soaked into the lot's old snow
and glass became the instant twin of ice—I just watched
with no remorse—

oh, that.

Maybe though I deny that sky-sent soil,
the stuff I never meant to catapult
that is, even now, still drifting down

I cannot lament this one
chilled instance, all my fault, again, spilled milk, and still—
for once, I quelled.

III. The Academy of Hay

Migration of Baling Twine

I have seen it in the truck bed, one fat spool
quiet as a salt block,
then rigged in the baler and sent
to bind each bale twice; I have grabbed it
like a man's suspenders yanking him to me,
Look here or sultry, *Sweetheart, where's my kiss?*

I have seen twine unhold with one touch
of knife, or in a bind his car key can
saw through, divided the packed bale splays;

I have seen these agricultural spaghetti
(orange plastic, dun burlap, fungicidal green)
drawn off thirty opened bales,
grouped in hanks, slumped over a nail
or hastily piled, as on slaughtering day we'll have
a heap of chicken heads, a heap of yellow feet,
here's a month's worth of yanked, chucked
twine plus milk filters, the newspaper, Dunkin' cups,
mastitis syringe, purple latex gloves shoved in
old grain bags for the dump from which

I have seen him take a bunch,
and deftly twist a halter for the calf; or link six to make a lead
or tie the milk-house door open so the flies can go,
one loop from knob to hook.

Flapping tarps quit with a quick pass-it-
through-the-grommet-cinch-it-down;
hoses hold their ovals with three snug wraps of a strand.

I have seen it eel its way, from one lump sum
to plenty of crude-ish sutures, like twenty extra fingers
pinching—as if the farm were a wound or a bird they keep trussed,

keep from blowing crawling falling growing away.

Part of an Argument

My girl tells me the droning in the sky
is the sound of our canted planet turning

on its ancient axis, and she thinks our hearts are pupae,
maybe three eons from flight. I'm afraid of both:

carrying to term; carrying interminably.
Now it's October, my last asters and tattered

cosmos drape over the rim, slant
away from their vase, their stems,

a galaxy
of axes.

Field Revealed as Runway by Morning

In case the pilot calls from his farm, *Hey, what are you doing?*
and in case you succumb, listening to his grin,
fourteen gauge wire bent with a hip into a crest
of mirth, in case this is distraction, as he tries to barter
your admiration for his impunity, chocks are stopping
the wheels of this engagement, and will stop,
or at least trip the rubber . . . well.
I am an unfenced field about to take delivery
of snow.

Hapless
(for Hannah Sullivan Randolph, 1923–2011)

I.

It's November. White moths dawdle
in my headlights like a broken strand
of snowflakes.

II.

Not long after she left us
all her amber marbles slid
into my décolleté, trickled to the sidewalk,
rolled into stunted grass.
                     Even later,
the last one sprung as I sat to pee, *plink*,
into the hotel toilet
                   just at the brink, yes,
I reached in as if to bring her back.

III.

It's twilight, and I've filled the wheelbarrow
full of pale cabbages.
Lifting the barrow's shafts, I push them
through dusk toward the char-dark barn, but
I might be wheeling her down the hall.

IV.

Four hundred miles, and forty days hence,
I want to ask as we jostle over the threshold,
Where would you like to go?

Winter as a Profit and Loss Statement

Saplings stuck in the bank will show the plow truck, in a month, road's edge.

The torrential rains, wind. Later, after dawn, clotted flakes cataract skylight.

Her finger traces the valley of his back.

The windows are opaque with ice, fissure lines squiggle like the seam where the plates of the human skull fuse.

Snow driven in the slice of space between the barn boards.

Snowflakes stick singly and doubly to a cow's roan coat, skewered to a hair.

To cool a hot pie they place it in the snow beyond the back door and cut into it an hour later. In the morning: the plate shape embossed there, ghost of the full moon.

*

The desk replaces the field, the lamp replaces the sun, one hand tills with a pencil.

She bangs the paddles of the gutter cleaner loose with a mallet.

Moisture beads on a cow's whiskers like dew on grass. The cow's bloated body is hard as a brick at the back of the barn.

She's felt a man go away without moving a muscle. She's seen a man veer away from a woman and he didn't move at all.

The chicken's struggle written in wing marks and scarlet beside the dog prints stabbed in the snow.

The plow's growl scrapes her out of a dream, dumps her back on the mattress.

Two songs: whine of the debarker at the mill, and crow's harsh caw.

*

Snowflakes: Are they heaven's search party?

Notes About Snow

The unexpurgated version
is actually immaculate.

My copy is also unabridged,
picture book: little lane, a bridge,

untold fields—a polemic, a manifesto of snow
utterly boring except for this neat strand of track:

someone's paw following its paw following its urge to take place
makes a dotted line I'm resigned

to sign, or
tear here.

One Wind

The cobweb trembling
an inaudible utterance
of something horrible
And the wind chimes

replying *it's fine,*
it's fine

The Present

The barn sits in the field like an opened gift amid the wrapping paper.

One morning after the towers fell, while we cleaned the barn
he grumbled, *It's not like it's a church.*

Sometimes snow on the roof blends into milk-white sky,
then the barn appears to have an open mind.

Remember those stone churches in the West of Ireland
where weather ravaged a thatched roof—

only cobbled walls and pane-less windows through which a bird might pass
into pews of ferns . . .

I kept hoeing splattered crap, ponds of urine into the trench;
he followed with a wide shovel, tossing sawdust across rubber mats

. . . and where the altar must have been, sheep bowing their heads to eat.

The Christmas Season

For example: the tinge of rancor from our squall
retreats, meek as sleet left in ruts and gulches,
and lawns revert to emerald, amenable.

We've improved our lot: hung earrings,
dabbed cologne, slung some colored lights
across the hay cart.

And now, seeing you in the forest—slight
exaggeration—the sudden grove at Aubuchon's,
a miracle: that we ever met. Hello.

I'm drunk on balsam's pungence
sprung in the air
where your shoulder nuzzles our tree.

IV. Husbandries

Her/Herd

My cow
her womb-shaped face:
her ready horns—anthers all
full of their pollens,

half-ton stand-in
for man and children.

Discovering Venus, April 1820

Despite what you've been told, the Milos

farmer does not find her in the ancient wall,

for he's been striding behind his mule,

using his share, turning the sod on its head—

surely, if the earth does anything,

it divulges everything: apricots, lima beans,

marble goddess. Brushing the soil

from her body, his mule waits in its trace.

He spits in his fingers to release

her features, eases the crumbs from her eyes,

the cool orb of her shoulder fits the socket

of his palm. He cannot divulge

everything, just as the earth keeps her

other portions, and after he doles grain

to the mule, perhaps, he stashes her

in the ancient wall and tells his wife

something of his day as night settles over

like treasure-littered soil.

On This Day in History

My crop is laundry; I reap it in awning-sized baskets.

Across the way, crows self-sow a field where

the Joneses' herd grazes between milkings.

I've got a premonition it's heavy with subdivision

even as Bessie, Dolly, and Holly, all Holsteins,

daughters of Aphrodite, cluster like the flies

pestering the edges of their very own eyes.

My crushes are pounded down at intervals

like fence posts: him-him-and-him—the wire's desire;

who suspects the heart abides like the hunk of beef sunk

in a grain bin before refrigeration's white mansions

got put in, pooped out, and were set out to pasture?

I fold a shirt, halve a sheet, not yearning, yawning; no fellow,

fallow, not even flagged, untrodden; stowed in the unsown oats.

First Do No Harm

The heart is a swarm of bees,
its queen smothered inside;
the cornfield is, in its other life,
an auditorium, the deer
—they all have walk-on parts.

All night in the Scotch pines at the edge,
the crows joke about men
with rifles, trucks, tractors—
the understudies; I mean, if I had to guess,
they know about abandoning the script,
about the massive unappeasable silences,

about pretending to be willing, rehearsing the end.
There were, easy, one hundred thousand leaves that fell,
like reminder notices, little cues. No one saw the moose
stagger through the corn; we just followed the bent stalks
into *an understanding*—we weren't nitpicky or petty,
just somehow not enlightened
enough.

Frightened enough
we are surrounded by creatures that meet
and mate and die with more finesse,
even if we don't meet,
even if the whole valley hears us.

In the Far Field and the Near

Out on the deck, a phoebe's clenching
top dowel on my laundry rack—

something I thought gone's come back—
this matter of early summer,

tossed into fall, like some crates toppled off
a truck by the blind curve,

beyond which another pickup lulls, two tires on the Reils' land,
man's hip against the fender, listening for his dogs'

yowls—nothing, but the breeze bang-slapping
the Atwoods' loose tin, which is god's way

of picking at a scab. Maybe Murray over in Stannard
sits zazen on these impulses, but I've got a hammer for that.

The Atwoods are devout Negligists, which is the discipline
of letting something busted take ten more years

to corrode, but not Dennis; he's rifle-ready with
wild apples piled in the clearing,

and in case you're angling for another chance,
I'm laying in wait for you here.

The Lasts

The last daylily, by the looks of it, has wept itself to sleep

                                                                 given everything,

I hate to watch it go: corn—ears, stalks, shocks—all's snipped into litter

                  la- la- la-   last hurrah,

season's bon vivant   of its own mauling   I'm no better—

dispatching three lambs into scraps   and tossing salt

(non-vivant) on their petal-soft   pelts

                                now dog robs mound for hock

now geese drain down to feed   on the least grains   of corn

spilled on the kill floor   of all of autumn—   this made bed

needs someone (a rural savant, a fall guy)   to lie

meanwhile the forest, by the looks of it, has amortized its stiffs wag

toe tags and on the third day the geese

(honky, itinerant, why shouldn't they?)   ascended

(semblance of a severed tie) into

                                                          elsewhere

Gust of Flowers Over Us

I love wind blitzing a blooming tree,
sending its petals apart

once, a gust confettied our windshield—
all I could think of was *blessing*;

as once his body stumbled backward,
and his hands stretched toward the ledge;

as once while clearing antique plates, slick
with au jus, my tray slipped, I reached,

as if I could halt anything, I once asked,
What if I fall in love with you?—

as I cleared the other tables
with the pieces in my apron clinking,

as they were working on the third story,
the scaffolding collapsed,

as he sees it, *I had to fall on my head to meet you;*
now he makes plates, of all things, for a living;

the cherry tree's in blossom—its mind made up
of petals—to bless is "to redden with blood,"

as the wind's having its way with them,
he explains, *They put me in a coma, they didn't know*

what they'd find when they brought me out,
they're cascading, bunches of blossoms,

the prettiest smithereens.

Kokoro

I know a woman who grows herbs in the cracks of her patio
so guests will crush them under loafers and swing-back flats—

sprigs of hyssop, oregano, thyme fume among the rum & Cokes,
herbs' generosity: this scent from contusion.

It is fair skunk reeks a mile beyond its struck body;
cells die, but odor only loses potency.

Who used to whiff dried apples before settling into write?
Like an orange thumbed open, torn from its rind, citrus sticking

to everything there's an art to how we abrade beneath each other's
needs, demands; if flesh absorbs touch and touch abides,

her ruptured greens scuffed underfoot
make spice a kind of grace.

Eating

She tweezes the sepal
from a honeysuckle
for its intense sweet bead

seated amid flange-like petals.
All of her attendant and
intent, all of her gathered,

a tiny fiber approaching
the needle's clean void
in the butter-cream

fragrant shade where
priest used to place host,
each honey bead sates,

and wakes,
her craving.

Forms of Worship

The Black Angus eat the bad batches, cracked batches
of flawed communion hosts the Sisters can not box and sell.

They eat gingerly, as if their tongues have been stung,
wafers dumped in a pile. What they don't eat is pushed

by hoof into muck black as themselves and the dark sacks nuns wear.
The uneaten hosts are buried, though I was taught as an acolyte not to let

a crumb escape. The fat priest would eat any bit left after communion.
All I do at the monastery is try to grow things: I place the tiny seed

in a mouth of dirt. I offer cheap beer in Dixie cups to discourage slugs.
When the imperfect body of Christ is crushed by the fantastic weight

of an Angus, my planted heart breaks
open: all I see are seeds.

Field Work

He shows me where to enter the field,
which direction to mow first—

then he gives me forty days of silence,
benign quiet, apart from the tractor,
a pasture where I can recall

all there was, aboard the wide mother-ship
winter, my first Quaker meeting,
all of us gathered, nothing said, aloud.

Later, in the same hay field, *Believe*
tracked out in boot prints: whoever
leapt into the letter doubled back

to make one part touch another
—both instrument and ink,
their whole self, written in snow,

not disappearing ink—disappearing paper.

V. Herd of the World

Sufficiency
September 11, 2006

You are holding their takeover tool, holding
the same simple implement of that morning,
five years later, but you are not using it on behalf
of nonstop ashes. Three hundred miles north, 7:20 a.m.,
you enter the field with it concealed in your vest
pocket, the cool fog hangs low, blindfolding
the road, there is only nearby: narrow dirt aisles
through the generous green rows, and the ruckus of
Canada geese descending with those notes
of panic, agitation in their unintelligible
commands. Your forefinger pushes the blade
out of its sheath, you hunch and begin sweeping
its edge through the slender necks of lettuce,
placing each head in the box you are dragging
along as the minutes go by quietly 8:45, 9:03,
9:59, 10:28; the lettuce stalk weeps a milk sap
on both ends—the part still rooted and the part freed;
dawn cruises into noon, you have everything you need.

Beforemaths, Aftermaths, and Polymaths
for K. Jeff Bickart

Somewhere the next deer
nibbles grave grass,
and the next you
stands in line at Ben Franklin.

The last deer, one you skinned
and hewed for a suit; and the last you,
the one who advised,
Always be finishing,

are still together in a frame—
hamming for the camera,
haloed in that coonskin
hat you trapped,

your flensing grin in tact:
invincible, entire—

what's left of your last
unravaged body.

o

I drew a finger through the dust,
feeding it white beans . . . a dotted line

and scooped a shallow dish, sprinkling
gherkin seeds—like flinging coins
to the hopper of the unmanned tollbooth,

will anything take?

The snapping turtle beside the road
swats a patch of dirt;

all summer here, on the mountain,
without electricity, under phases of the moon,
a famous writer fathers essays;

we're sitting at his picnic bench
when I feel a twinge
near my right kidney—

little grain of me or star,
more silica than cell,

you sort of something,
you begin your

fall toward loam.

(G)loves
"With his work, as with a glove, a [wo]man feels the universe."
—Tomas Tranströmer

of planting, such as when we tuck each seed potato in the earth,
a hand-span apart, burying them two fingers deep, until we've put
a bushel underground, where they remain interred until the chilly
months that follow, when we reach back into the dirt and twist the
numerous doorknob tubers loose

of measuring, as the word for the Greek coin drachma means
"a handful," so I stride into the newly plowed place, strewing handfuls
of buckwheat seed from my paper sack, an extravagant act, like casting
pearls. Then I tally the farm with my limbs, as each row is eighty-five
of my footpaces. And nearby the broccoli starts go in the ground
three-quarters of my forearm apart. In another bed I make a fist:
two knuckles for every parsnip

of grasping, as when the cat sets a blue bird loose in my bedroom and
I recapture it, closing my fingers around the bulb of blue feathers,
absorbing its scampering pulse. Releasing it to the cherry tree in the yard,
then my hands seem hollow

of harvesting, as when I plunge my hands and cinch fingers around
loosened carrots and draw them to light, wrench off their feathery tops,
then drag the bulging sack of them back to cellar

of ravaging, as when butchering our poultry requires tearing, shucking the feathers back like husking corn, fistful after fistful, and afterward, my hands are scalded and ringing, as after a long fit of applauding

of caressing, as when he fills my teacup, then clamps and kneads my shoulders, then pours a small pond of lotion in his palm. He rubs his hands together, and now takes hold of mine, one at a time, working lotion into the creases, thumbing it into my shovel-handle calluses

of asserting, as once while we were transplanting, I asked the new girl what tool she used most often on the soil, and she sidled up to me slammed her hands down on the bed, "These!"

which resemble lily petals opening into the wide palm of welcome

or fallen oak leaves, blind, feeling the face of the earth

or a flapping crow whose wings are glossy black gloves

And then after a handful of years I was no longer mesmerized by gloves wherever they dropped like stunned birds. Instead I studied these unsheathed instruments whose sowing and reaping, seemed from a distance, seemed like the act

of stooping to shake hands with the earth.

The Needle

I teeter
between extremes—

one hand finishes the beetle
sleeping by his leaf meal,

the other plucks you a raspberry;
this hand cups an asterisk of chick:

later this hand cradles the axe that lops
its matured head, an apostrophe

this hand tosses in a bucket,
this hand lifts the fork:

Judas and Jesus at the ends of my arms
are both full, so your letter's clamped

between my teeth, as I try not
to bite, to kiss.

Twine

He wreathes a piece of twine,
secures it around her neck and holds
the tail of it in his work-fattened hand.
He offers her the stub of his thumb,
which she takes for a teat. Stretching
her muzzle, her taut legs step toward him
at the moment he takes one step back.

This is how they progress:
the one-day-old calf on legs about to buckle,
baited with an outstretched thumb; an odd waltz:
he recedes at the rate of her pursuit,
until they've crossed the road as a couple,
where his wife waits by the open gate
as if she has not seen this before.

The Heifer

Eats like she's painting with her nose—
draws with a back-and-forth motion, not on canvas, hay.

The cat licks his flank: strokes but no color.
My brush is a broom; I love to sweep—

my primitive urge: not to eat, nor slake, but purge.
To search the disbursed for reunion.

With a bit of spit on my fingers I could kneel
by the pan and fashion

debris of the last catastrophe into
a whole woman, a mended man.

Two Eggs

This one the color
of my shoulder in winter,
and this one, my shoulder in summer.

No seam no pock no
porthole, smooth as oil.

The surface curve:
just a tip and a buttock,

silent as a horn in the trunk,
how many times can we give

what's formed inside us—
Never? Always? Once?

Nine Miners

I.
The wall broke and water shot out.
They all called to God
to have mercy, and the water stopped.

Above them, a farmer
ordered an air hole
drilled through his pasture.

Meanwhile they tied themselves in a gang line,
a collective decision to live or be recovered
as man with thirty-six limbs;

Even if I invented this, it's true?
Part of us is created empty
and yawns to be filled.

My longing's gaped like
the wheelbarrow,
and the ash pail,

the laundry basket, the coffee mug,
the stock pot and dough bowl,
the fry pan darker than a pupil,

an anthill, an air hole . . .

II.
I've got a rock bar—an iron javelin,
a stupendous pen for (de)scribing
the cavity—I let the weight do the work

of making a vacancy as I heft and let it plummet,
its tip drills a mouse hole into a gopher hole
until the bar itself is buried up to its hip,

if it had hips. I stir, or maybe I row, the opening.

III.
Here's my final example: the swallow home
with its inch-wide void
into which
        one sleek one
goes like a morsel over the gullet—

when they fledge
a gorgeous disgorgement—
one two—
        three . . .
. . . six seven

                and finally, nine
                        (over time)
come to light.

Mappa Mundi

God in
particulate;
god as crumb;
god as immense friable matrix
of granulated mountain, deceased faces, feces, godlettes.

Suppose
the sons of god
addle a goddess—anhydrous ammonia,
and drag the chalice edge, called *speed the plow*,
through the body of god, for the bread of humankind, for a side of beef.

I don't believe it.
I don't think: oh, holy soil.
Dirt—it doesn't even have eyes for chrissake,
even if I'm a part of earth before, now, evermore,
like the river testifying as river from its midway station as rain.

The Holding Pen

The way you swipe up scattered crumbs
is the way you steer the herd toward the barn
(your cupped palm the stockyard, your sponge-hand giddyaps)

and the way you rustle hours from the given day,
driving them out at daybreak, and back again at dusk,
all those unbranded moments rushing through memory's chute,

and the way you boot any egregious behavior out of the pasture,
where it fattened on germane complacency,
and shoo it into the dark slaughterhouse of *Hell No*

partners the way you're bound to accept the lumps as sum—
be they foundered horses or broken estate.
Lone star of Shorthand Ranch? Believe in piecemeal.

The way the priest treats even a broken host as whole.

An Exegesis

She yanks bales off the hay elevator and thrusts them

toward the farmer who stacks them to the rafters,

charging the barn's imagination with soil's offspring,

fifty-five acres of timothy, nothing but the next bale

and the next bale and the next bale; she can't see

the others below, unloading the cart or guess

how far they have to go; *on the third day*

he rose from the dead and ascended into heaven . . . where she

grips him by the twine, hoists high, heaves him over.

Timothy

When one grass head
casts its shadow:
bear fur, the boy's nape
your mane, your hay field leaning,
toward the woods.

About the Author

photo / Howard Greene

Julia Shipley is an independent journalist, farminista, and the author of *Adam's Mark: Writing from the Ox House* (Plowboy Press, 2014), selected as a Boston Globe Best Book of 2014. Her writing has also appeared in *Alimentum: The Literature of Food, Burnside Review, CutBank, Cincinnati Review, Colorado Review, December Magazine, FIELD, Flyway, Fourth Genre, Green Mountains Review, North American Review, Orion, Poetry, Poet Lore, The Rumpus, Small Farmer's Journal, Taproot,* Terrain.org, *Whole Terrain,* and *Wildbranch: An Anthology of Nature, Environmental, and Place-Based Writing.* A two-time recipient of grants from the Vermont Community Fund and the Vermont Arts Council, she's also been a fellow at The Frost Place and The Studios of Key West. A native of suburban southeastern Pennsylvania, she spent her twenties migrating from farm to farm working for Community Supported Agriculture projects in the Northeast until 2004 when she began Chickadee Farm. She is married to one man and six acres in Vermont's Northeast Kingdom.

About Bona Fide Books

The Melissa Lanitis Gregory Poetry Prize was created in memory of Lake Tahoe artist Melissa Gregory, the inspiration for Bona Fide Books. The prize is awarded annually for an unpublished collection of poetry. The winner receives a cash prize, publication, and a reading at Lake Tahoe.

Bona Fide Books is a small press on a mountaintop connecting writers with their readers.

www.bonafidebooks.com

This book is set in Baskerville, Adobe Caslon Pro, and Optima, and it is printed on 30 percent postconsumer recycled paper, processed chlorine-free.